# GRINDING, LAPPING AND HONING

*The application of abrasive processes to finishing surfaces to a high degree of precision, with descriptions of machines from the operator's viewpoint*

*By*

EDGAR T. WESTBURY

First published 1950
Reprinted 1987
TEE Publishing © ISBN 0 905100 68 9

## CONTENTS

## CHAPTER I

# INTRODUCTION

Abrasive methods of working materials have been in use ever since the world began; indeed, it is reasonable to suppose that some of the earliest attempts to shape natural materials to human ends involved crude methods of grinding, lapping and honing. Long before the existence of cutting tools, primitive man shaped his edged weapons—flint axes, spear and arrow heads—by chipping them roughly to shape and rubbing them smooth and sharp on flat pieces of sandstone, or other stones having abrasive properties. At a much later date, when it became possible to smelt and forge metals, abrasive methods, improved in detail but exactly the same in principle, were used for polishing all kinds of metal objects, and also for sharpening edged tools and weapons.

For hundreds of years no further function of abrasive principles was visualised, but improved means of applying them gradually developed, first in the shape of the rotary grindstone, and afterwards in the lapidary's wheel, which developed later into the modern lap and the polishing lathe. Eventually, the discovery of highly efficient abrasives, both natural and artificial, and the development of high speed, led to the realisation not only that many processes in the working of metals and other materials could be ***efficiently*** carried out by abrasives, but that the rate and precision of the operations could be very finely controlled. At this period, the engineering industry had begun to realise the value of precise limits of dimensional and geometric accuracy, and for this reason, coupled with the fact that abrasive methods could be applied to the hardest metals with practically the same facility as to those of a more tractable nature, means were soon found of adapting them to various machine tools, for use on work requiring special accuracy.

Apart from tool grinders of various types, most of them fairly simple, but some elaborately equipped for dealing with special cutters, the most popular and important machine in the abrasive class is the cylindrical grinder. This was first developed as a "tool room" machine, applied to few jobs outside the production of hardened gauges, mandrels, and so on; but it became an essential production tool as soon as the practice of using hardened shafts and similar wearing components in engines and other machines became general. This was because the hardening of the shafts caused distortion and scaling, which destroyed their initial accuracy; and the only means of correcting this was by leaving them oversize, to be finished and corrected afterwards by grinding.

It was soon found, however, that grinding processes could be advantageously applied to the accurate production, at commercially-practicable rates of output, of all sorts of work, whether hardened or otherwise; and although nowadays it is possible to harden work with little risk of either distortion, or scaling, the number of components now finished by grinding is greater than ever. Grinding has been applied to practically every kind of machine tool operation with substantial improvement in finish and precision.

But grinding machines, important as they are in modern engineering, are by no means the only important application of abrasive processes in engineering industry. Within recent years, considerable developments have taken place in lapping and honing processes, which have been found particularly applicable to the production of highly finished and accurate wearing surfaces, and special machines for applying these processes have been introduced. Thus the most antediluvian and primitive metal-working processes have found a useful place among the most up-to-date methods in modern practice.

This handbook is intended to describe, in a practical way, the various methods whereby abrasive processes may be used in engineering practice, as a means of finishing surfaces to a high degree of precision. Its scope is not extended to methods of surface finish in which the sole object is to improve *appearance*—in other words, abrasive polishing—as it is considered that this subject calls for special and entirely separate treatment.

It is inevitable that, in describing the principles and practice of abrasive processes, a very large proportion of the book should be devoted to the various types of machines used in these processes. In modern practice, the machine is an essential factor in the control of abrasive action, and certainly could not be dissociated from any consideration of the precision which is the principal reason for the use of these processes in engineering. The machines are, however, described from the point of view of what they will do and how they do it, rather than by mere catalogue specification ; and in describing their manipulation, emphasis is laid not only on the *how*, but also the *why*, of the various operations.

## The three processes

The abrasive processes defined respectively as grinding, lapping, and honing, are described in detail in their appropriate sections of the book ; " grinding " is given by far the greatest prominence, because it has by far the widest application in engineering, but the inclusion of lapping and honing in one chapter at the end of the book does not mean that they are relegated to a position of minor importance.

It is appropriate, in this introduction to the subject, to define briefly what is meant by the terms grinding, lapping, and honing, in order that the reader may be under no misapprehensions as to the purpose and scope of these processes. Brief definitions can only be general and incomplete, and may, therefore, be dangerous ; but as more fully detailed definitions are given later, it may be stated that grinding is understood to mean the use of solid abrasive wheels to shape or finish material by removing greater or less amounts of material ; lapping and honing are both fine shaping and finishing processes in which very small amounts of material are removed by a combination of reciprocating and rotary motion, the former by the action of loose abrasive particles and the latter by using solid bonded abrasive sticks. Either lapping or honing may be used for superfinishing work which has previously been ground, or otherwise machined.

The examples of grinding machines included in this book are mostly of British manufacture, and without exception, are known to be capable of high class precision work, but it is not claimed or implied that no other makes of machines in the respective classes are capable of equally good work, and no comparisons are drawn between the merits of features of design and construction of the different machines. It is fairly safe to infer that there is no room in the modern engineering factory for machines which will not carry out the work required of them efficiently, expeditiously and precisely.

Due acknowledgments are made to the many machine tool makers who have assisted, with data and illustrations, in the preparation of this book, and whose names are mentioned herein. It is hoped that the pains taken to obtain the most up-to-date practical information possible, on all matters dealt with, will enable the book to fulfil the requirements of modern machinists and engineering students who wish to obtain a comprehensive knowledge of these vitally important processes.

## CHAPTER II

# PRINCIPLES OF GRINDING AND ABRASIVE PROCESSES

THE action of abrasives is often regarded as a mere wearing away of the surface of the material, the particles which are removed being mere impalpable dust. Although this may be more or less true of some of the older abrasive processes, modern abrasives have a definite cutting action, as can easily be proved by examining the sludge from a grinding machine under a low-power microscope ; indeed, the evidence of the naked eye is sufficient, in cases where wheels of coarse grit are efficiently used.

Modern grinding wheels may be regarded as rotary cutting tools having an infinitely large number of cutting edges, each formed by a tiny fragment or crystal of very hard material, imbedded in a matrix of softer material. Individually, these cutting edges are capable of removing only a very small amount of material from the work being ground, but owing to the large number of them which can be brought into action in a given time, the wheel is capable of removing the material at a fairly efficient rate, depending upon its size, peripheral speed, and the way in which its particular properties are utilised.

In a machine tool using single-point cutters, such as a lathe or a planing machine, the force exerted at the point of the tool is considerable, especially when a fairly heavy cut is taken at low speed; and although, within certain limits, speeding up the lathe and reducing the depth of cut will reduce the load without reducing the amount of metal removed, it is always fairly heavy. But although the maximum force is exerted against the cutting face of the tool, it is also necessary to exert a fairly heavy contra-radial force to prevent the tool springing away sideways from the work, and thus the structural parts of such machine tools must be extremely strong and rigid.

By increasing the number of cutting points in action at once, the load on each point may be considerably reduced, and although this does not reduce the power necessary to drive the machine for a given rate of metal removal, it does reduce the side force necessary to keep the tools in action, and there is less tendency for inaccuracy to be set up by springing of structural parts. Milling machines and other rotary tool machines exploit this principle to some extent, but not in any way approaching the action of the grinding wheel, where the side pressure required to keep the work and the wheel in cutting contact is extremely small. The higher the speed of the wheel, the more cutting edges will be brought into action in a given time, and the lighter will be the load on each of them for a given rate of stock removal; but there are other considerations which govern the wheel speed, as will be seen later.

In lapping or honing processes, the same principle of using a very large number of cutting edges is applied, but in a modified manner; the action of the abrasive crystals is slower, and they are usually more heavily loaded, so that these methods are better suited to removing very small amounts of metal, under conditions best suited to close control. This applies also to abrasives used for scouring and polishing, but as this subject is scarcely within the scope of this book, it is not proposed to pursue it further.

## Abrasive Materials

Before going further into the action of abrasives, it is advisable to devote some time to considering their nature. The early abrasives were in all cases natural substances, used either in solid or finely divided form. As already observed, sand and sandstone have been very extensively used, and to a certain extent are still used ; many craftsmen hold that the slow-speed sandstone is unequalled for grinding keen edged tools, owing to its mild action.

Nearly all the oxides of silicon and aluminium, and to some extent those of magnesium, also their chemical and mechanical combinations, have some virtue as abrasives, but where they are found in a natural state, the presence of impurities affects their efficiency as grinding wheels. The best-known natural abrasives are emery, corundum and garnet, but all these are regarded as more useful for coating abrasive strips and sticks nowadays than for use in solid or bonded form. Corundum is an oxide of aluminium having extremely hard crystals, and would be very valuable as an abrasive, but for the fact that it usually contains a substantial percentage of useless impurities, which impair cutting, increase friction, and tend to cause clogging. Emery is a natural mixture of corundum with various iron oxides, which also are worse than useless in respect of abrasive properties.

By far the majority of abrasive processes at the present day are carried out with artificial abrasives, the composition and purity of which can be very closely controlled, so that their efficiency, and also their suitability for particular purposes, can be guaranteed.

One of the most useful of these abrasives is carborundum, which was discovered by Dr. Edward G. Acheson in 1891. It is a carbide of silicon, made by fusing in an electric furnace a mixture of coke, sand, sawdust and salt in carefully regulated proportions. The resultant product is a mass of tiny acute-angled crystals, of a hardness second only to that of the diamond—in short, something closely approaching the ideal abrasive material. This mass is reduced, by crushing, to a powder, and sifted through sieves of different mesh to separate and select the different sizes of grain or "grit," which are subsequently graded by numbers which correspond to the mesh of the screen through which they will pass. Thus, to define a wheel as " 50 grit " means that the individual particles of abrasive of which it is composed are capable of passing through a sieve having 50 meshes per linear inch.

The extremely hard aluminium oxide crystals, which occur naturally in corundum and emery, are also prepared artificially, by heating the aluminous clay known as bauxite (which is also the raw material used in making metallic aluminium) in an electric furnace, so that it is completely dehydrated and assumes a sharp crystalline structure. This abrasive is known commercially as " Aloxite." Other artificial abrasives are " Alundum " " Adamite," and several others classified under special trade names ; all have their own particular properties and uses.

Of recent years, the extremely hard alloys used for cutting tools have presented special problems to the grinding wheels for sharpening them, as the hardest and sharpest abrasives in normal use have sometimes been found incapable of grinding them efficiently. A special brand of carborundum, known as " Green-Grit," which is exceptionally hard and sharp, has been found suitable for this purpose, but in many cases the hardest of all known materials—the diamond —is used for making grinding wheels for these special alloys. Either natural or artificial diamonds, crushed and screened to select the grit size, are used in a manner similar to the more common abrasives. Diamonds, which are a crystalline form of pure carbon, have been used in a powdered form for lapping purposes for many generations, by jewellers and lapidaries.

### Bonding

Grinding wheels are built up from the selected grit by mixing it with a plastic material and moulding into suitable shape. Various kinds of cements are used as a bond, according to the purpose for which the wheels have to be used. Vitrified wheels have a clay bond, and are fired in a kiln at high temperature, so that the clay hardens into porcelain or its equivalent. This forms one of the strongest possible bonds, and also has the advantage that the slight contraction of the clay when fired leaves the wheel slightly porous ; thus it can be more effectively cooled by flushing with water during use, than a solid and impervious wheel. Silicate wheels are bonded with sodium silicate (waterglass) and are particularly distinguished by their mild cutting action, which makes them suitable for sharpening edge tools ; they should not be used for rough or heavy work.

Shellac and other organic gums are used for bonding wheels for special purposes, but these wheels are generally used dry, as they are impermeable to cooling fluid. Rubber is a useful bond for light and delicate wheels used for cutting off tool steel bar and similar purposes, as it imparts a certain degree of flexibility to the wheel. An even better bond, which gives the highest possible mechanical strength, is obtained by using synthetic resin (bakelite), and the " Redmanol " series of wheels of this type introduced in recent years by the Carborundum Co., is suited for cutting off at surface speeds as high as 16,000 ft. per minute.

## Grade

This quality is defined as the degree of tenacity with which the particles of abrasive are held together by the bond, and is thus a measure of the force required to tear them from the wheel under the effect of heavy grinding load. The various grades of grinding wheels are designated by letters of the alphabet, from hard to soft. Carborundum and "Aloxite" wheels are graded from D to W, those from F to H being grouped in the " hard " range, from I to M in the " medium " range, and N to T in the " soft " range. Letters outside these ranges refer to extra specially hard or soft wheels, which are not often used in normal practice.

It is important to remember that the grade or " hardness " of a wheel does not refer to the hardness of the abrasive particles, and that the term "grading," is not the same as grade, but refers to a system whereby all the essential properties of a wheel can be specified or defined by a code symbol of letters and numbers, as compiled by the maker or group of makers of certain wheels, and fully explained in their specification lists.

## Effect of Grit and Grade

If we return to our original analogy and regard the grinding wheel as a milling cutter with an infinite number of cutting teeth, a little further consideration will show that there is one great and irreconcilable difference between them which is not properly accounted for. A milling cutter, when it becomes blunted after considerable use, can be, and must be, re-sharpened to bring it back to its original efficiency but it is clearly impossible to sharpen each cutting particle of a grinding wheel.

In practice, as the particles gradually become dulled, the force which must be exerted to make them cut increases, until it eventually overcomes the tenacity of the bond, and the particles are torn away from the surface and discarded; the surface of the bonding material also wears away, bringing into action new and sharp particles beneath. It is necessary to select wheels for particular purposes so that the rate at which the particles of abrasive are torn away is just sufficiently fast to maintain the cutting efficiency, and no more; also so that the grit is neither too coarse nor too fine for efficient removal of metal at the required rate. The proper selection of wheels is, however, a matter of practical experience rather than theoretical reasoning, as there are many factors which modify the behaviour and qualities of wheels. The makers of abrasives have taken the utmost pains to gather and classify all available information on their use, and their advice is practically infallible for all normal work. On no account should the importance of selecting the correct grit and grade of wheel be underestimated, for it is one of the most vital factors in efficient and precise work.

## Trueing and Dressing Wheels

These two terms are often regarded as meaning the same thing, but this is not correct, although the processes may be combined and simultaneous. Trueing consists of correcting the concentricity or contour of the cutting face or faces of the wheel ; dressing consists of cutting the wheel face or faces to remove clogged particles and improve cutting action

It is most essential that wheels should run perfectly truly, not only to ensure regular cutting action, but also to ensure perfect balance. Dressing is necessary at intervals when the wheels become dulled or loaded with metal, but it should not be too freely indulged in, as it reduces the size of the wheel, wastes material, and reduces surface speed.

The most useful and efficient dressing tool is a diamond tool, mounted so that it can be traversed across the wheel

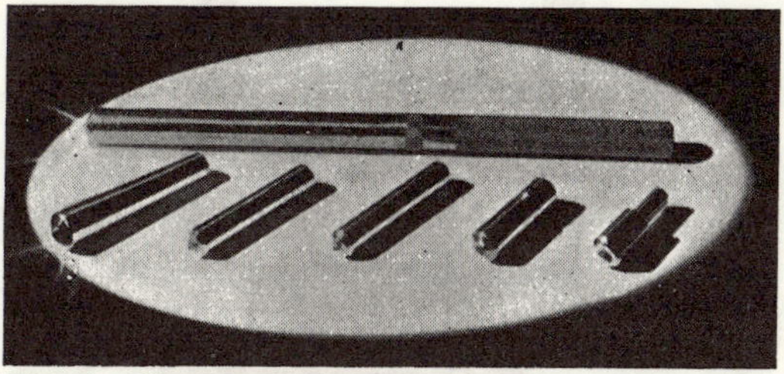

**Fig. 1. Group of diamond tools for trueing and dressing grinding wheels.**
(*Alfred Herbert, Ltd.*)

face by the slide of the machine. Tool grinders can be fairly efficiently dressed by the " star wheel " type of dresser, which removes the wheel particles by percussion ; hard abrasive slips, and in some cases rotating abrasive wheels, are also used for trueing and dressing .

Diamonds for wheel trueing are usually set in the end of a steel rod which may be furnished with a handle, or adapted to be mounted in a slide-rest fixture. A group of diamond tools is shown in Fig. 1.

## CHAPTER III

# TYPES OF GRINDING MACHINES

THE most common and essential type of grinder is the ordinary tool grinder (Fig. 2) for freehand, or as it is generaally termed, " offhand " grinding of various small tools, in particular, ordinary single-point lathe tools and cutters. It is an indispensable tool in every engineering shop, and may, with intelligent use, be applied to a very wide range of work

**Fig. 2.—Typical example of a double-ended tool-grinder**
*(S. Tyzack and Son, Ltd.)*

with reasonable efficiency, but it must be confessed that it is more often abused than properly used. Wheels of incorrect specification, run at incorrect speeds and grossly overloaded, are all too common, even in machine shops quite efficiently organised in other respects. For these sins, the blame must be laid on the users of the machines, as their makers have done everything possible to ensure their efficient performance, and the modern examples of tool grinders, however simple in form, have very carefully designed and constructed bearings and spindles, and provision for efficient drive.

Following the tendency which has long been prevalent in other machine tools, many modern tool grinders are equipped with individual motor drive. In some cases the motor is actually on the wheel spindle, or, to put it in another way, the grinding wheel or wheels are mounted on the shaft of the driving motor.

The increasing use of multi-point tools, such as machine reamers, rose-bits, and milling cutters, also the need of greater accuracy in grinding twist drills and ordinary lathe tools, made it necessary to introduce work-holding fixtures by means of which the tool to be ground could be presented at exactly the correct angle to the wheel, and the extent of the grinding properly controlled. Some of these devices have been made in the form of attachments to plain grinding machines, but it has generally been found more desirable to incorporate them in specially designed machines. The ultimate result of this development is the modern universal tool and cutter grinder (Fig. 3.) The use of a twist drill grinding attachment, or an inclinable table to facilitate accurate angle grinding of lathe tools, on fairly simple and inexpensive machines, is, however, fairly popular.

Machines designed specially for cutter grinding always have a slide on which can be mounted a pair of headstocks, with centres for mounting mandrels or cutter shanks, to enable them to be truly revolved; often one of the headstocks is equipped with some form

**Fig. 3.—The "Alba" universal tool and cutter grinder**
*(B. Elliott and Co. Ltd.)*

of power drive. The machines thus become virtually capable of carrying out cylindrical grinding, and although not specifically intended for work of this nature, are quite useful for small jobs within their capacity.

The machines designed specially for external and internal grinding, however, are generally much more massive in their structure than tool and cutter grinders, with the object of ensuring the utmost steadiness under working conditions, and avoiding the slightest deflection of either the wheel head or the work-holding headstocks. Some machines, such as that shown in Fig. 4, are made for external grinding only, others for internal work; others can be adapted for either purpose by the use of a dual-spindle wheel head, or changing the essential parts.

Although cylindrical grinders handle a good deal of the work formerly carried out by lathes, and may be said to be a modern evolution of lathe principles, they differ greatly from the orthodox types of lathes in their design, The bed, which is massive and stationary in most lathes, is of lighter construction, and mounted on a sliding carriage in grinding machines, and the wheel head, corresponding virtually to

**Fig. 4.—Cylindrical grinding machine for external work**
*(Newall Engineering Co. Ltd.)*

the tool slide of a lathe, is solidly mounted, and not usually designed to have more than a very limited movement.

Similar principles are incorporated in the design of surface grinders, which may be regarded as developed from milling

**Fig. 5.—"Precimax" vertical-spindle surface grinder**
(*John Lund, Ltd.*)

machines, and are made with both horizontal and vertical spindles in a wide variety of types to suit different classes of work. A typical vertical grinding machine is shown in Fig. 5.

At the present day, practically every kind of machine tool which uses cutting tools has its counterpart in some form of grinding machine, with the possible exception of the drilling machine. There is no particular object in attempting to drill holes from the solid by grinding methods, which would be relatively inefficient, because of the impossibility of running the wheel at sufficiently high speed to cut properly at the centre. Drilled holes can, however, be finished by internal grinding when special accuracy is required, and in tool room practice quite small holes, down to $\frac{1}{4}$ in. or even smaller, are thus dealt with.

The principle of the horizontal boring machine is adapted to grinding in certain types of cylinder grinding machines, which have a "planetary" grinding wheel carried in an extended rotating head, and adjustable to suit the bore of the work, which is held stationary on a sliding work-table.

The principle of the planer is also adapted for grinding, with horizontal, vertical, or in some cases, angular spindles; and other machines are used for the continuous grinding of flat surfaces on gangs of small pieces secured to a horizontal rotating table. Grinding processes are also steadily gaining favour for finishing splines, gear teeth and worms or screw threads.

Many types of portable or semi-portable grinders can be applied to lathes and other machine tools, or to facilitate hand operations. Both types range from very small to quite large sizes, and are in many cases equipped with electric motor drive. Alternatively, grinding heads driven by flexible shafting or by belting are also applied to various purposes. The machine tool grinding head attachments have evolved from the "tool post" grinder, which was originally developed for the purpose of trueing up lathe centres or for carrying out simple and occasional internal or external grinding operations. Attachments of this type, an example of which is seen in Fig. 6, are often clamped to the spindle head of a milling machine, or the ram of a shaper, and used for surface grinding.

Many motor-driven grinders intended primarily for hand use are also capable of being clamped to the tool-post of a lathe, or on other machine tools. Flexible-shaft driven grinding heads, obtaining power either from the lineshaft or an independent motor, are extensively used for dressing large castings, and similar work which is too cumbersome to be readily applied by hand to the surface of an ordinary grinding wheel; an example of taking the machine to the work, instead of the work to the machine.

Pneumatically-driven grinders are similarly applied as machined tool attachments or hand tools, and an interesting example of this type of machine is the Desoutter turbine-driven grinder, which attains incredibly high speed, and is intended particularly for dressing dies and press tools with tiny " button " grinding wheels.

Large grinding heads, equipped with electric or compressed air motors, and in many cases fitted to a carriage or compound slide, are used for squaring up the ends of girders, railway and tramway lines, or for dressing welded joints in them. Commutators of large electric motors and dynamos may be trued up *in situ* by means of grinders similar in essentials to those used for attachment to the tool post of a lathe.

The utility of grinding attachments to machine tools is beyond all question, but it is a mistake to regard them as adequate substitutes for specially designed grinding

**Fig. 6.—" Dumore " electrically driven tool post grinder, performing internal grinding operation in lathe**

machines. Their real merit lies in their convenience, rather than their accuracy or efficiency, and so long as this fact is fully realised, and they are not expected to perform tasks entirely outside their capacity, they can be made to fulfil extremely useful purposes in every engineering workshop.

## CHAPTER IV

# TOOL AND CUTTER GRINDERS

THE old-fashioned "grindstone" or slow-speed sandstone wheel, which was for very many years the only method available for grinding either woodworking or metal working tools, can scarcely be dignified by the name of "machine." In many cases the stones were mounted on roughly forged shafts, in bearings of the most rudimentary kind, and housed in a wooden frame or trough, the latter being used to contain sufficient water to immerse the edge of the stone. A sandstone must be kept wet while in use, or its surface will rapidly fill and glaze so as to lose its abrasive power almost entirely. The water trough, however, was a prolific cause of deterioration of the stones, because if they were left dipping in water while stationary, the local absorption of water by the porous stones would cause a variation in their hardness, so that they wore badly out of truth. A constant water feed service to the wheel, and means of draining it away from the trough, is a much better arrangement. Sandstones are still used for grinding hand tools in woodworking industries, and their utility can be much improved by arranging them on the lines of modern grinding appliances, but they are essentially slow-speed wheels, and apart from other considerations, their low tensile strength would not allow them to be run fast enough to compete with efficient modern abrasive wheels.

High-speed tool grinders necessarily embody at least two mechanical improvements—accurate spindles and substantial bearings—and their design follows the principles of the polishing lathe rather than the grindstone. In fact, light machines for jobbing work often embody provision for mounting either grinding wheels or polishing mops. Another feature which is generally regarded as essential to these grinders is a tool rest which enables some measure of control to be exerted over the angle at which the tool is presented to the wheel. In many cases, some method of partly enclosing the wheel is provided as a measure of safety against flying dust and sparks, or more serious emergencies ; and allowance may also be made for water service. Water cooling is in nearly all cases a great advantage, as it allows the wheel to be used efficiently, with the minimum risk of drawing the temper of the tool, and results in finer finish with a wheel of a given grit.

It is most essential that the spindle of a high-speed grinder should run truly, and also very freely in its bearings, but without undue play. The wheels must also be mounted so that they run truly and in proper balance. It is usual to bush the centres of wheels with lead to fit the size of shaft on which they are intended to be mounted, and in the event of the wheels having to be used on different sized shafts, the bush should either be bored out quite truly, or fitted with a concentric liner bush machined to close internal and external limits. The wheels are almost invariably clamped by side pressure between washers of large diameter, and it is advisable to interpose washers of resilient material, such as blotting paper, between these and the wheel faces, in order to ensure perfectly even pressure and avoid local strains. The nuts used for clamping the wheels are tapped right and left hand for the respective ends of the spindle, so that they tighten up in the direction opposite to that of the rotation ; thus they cannot be loosened by forces due to load or inertia.

## Inclinable Work Table

The ordinary tool rest fitted to most plain grinders is not really adequate for controlling the angle at which the tools are ground. In many cases only a narrow ledge is provided, mounted on a stalk, which is clamped in a socket attached to the machine bedplate, and is thus capable of limited adjustment in a vertical direction only. By this means, the equivalent of an angular adjustment is obtainable, as the surface of the table is raised or lowered relatively to the horizontal centre line of the wheel; but this applies only when grinding tools on the peripheral face of the latter. Even this elementary method of angle control is not always utilised to its full advantage.

**Fig. 7.—The "Milford" motor-driven tool grinder**
*(B. Elliott and Co. Ltd.)*

The value of the exact observance of proper rake and clearance angles on cutting tools is rapidly becoming more and more fully realised, and this indicates the importance of really positive and reliable angle control on tool grinders. Machines, specially equipped for this purpose, are to be found in most modern tool rooms, an example being shown in Figs. 7 & 8. It is intended for either wet or dry grinding, and is equipped with large work tables inclinable on either side of the horizontal plane, and also equipped with sliding guides which can also be swivelled so as to control the nose angle of the tools. This device is adapted for use with annular or cup wheels in which the side face is used for grinding in preference to the peripheral face. This ensures that the ground face shall be dead straight, and thus of constant angle; which is not the case in tools ground to a concave arc on the periphery of a wheel. It is in this way much easier to verify and ensure the maintenance of the true cutting angle. Concave or "hollow" grinding has its own particular merits in certain cases, and sometimes special small diameter wheels are used in order to accentuate the concavity, when grinding the raked faces of certain tools, in order to encourage the tight rolling and easy clearance of the swarf.

A flat work table is useful only when the tools are dead straight; thus forged swan-neck or "dropped" tools, or cutters held at special angles in holders, require special treatment. It is nearly always possible to mount these in special holders or jigs with flat bases to fit on the work table, but some machines, such as that shown in Fig. 9, have been equipped with universal slide-rest controlled holders which can be adjusted to suit any type of tool.

The side faces of plain disc wheels should not be used more than is necessary, because the considerable difference in the peripheral speed at different parts of the face makes it impossible to ensure the most efficient speed for the work. Wheels intended for side grinding always have a comparatively narrow overhung face, which may readily be trued to perfect flatness after wear takes place. Some grinding heads, however, are not suited to withstand the end thrust which invariably takes place when using the side faces of wheels.

**Fig. 8.—Inclinable work table and angle fence of "Milford" grinder**

**Fig. 9.—Lumsden oscillating tool grinder**
*(Alfred Herbert, Ltd.)*

Grinders intended for wet grinding are generally equipped with a power-driven pump and circulating system for the coolant, and the heavy duty of grinding high-production tools demands that a copious supply of fluid be constantly maintained.

## Twist Drill Grinders

Many devices have been introduced to control the tip and clearance angle when grinding twist drills, not only to ensure that these angles are correct for efficient cutting, but also that the tips are ground symmetrically, so that they share the work equally and that the drill will cut to the correct size. These devices may be fitted as an attachment to a plain grinder, or incorporated in the design of a special machine (Fig. 10). Several different principles of operation are used in the various appliances, which are all capable of giving good results if properly handled. The most difficult problem in any form of twist drill grinding appliance is that of setting the tip of the drill in its jig or holder so that it is "indexed" correctly relative to the flutes. This position cannot be settled by reference to a fixed facet or tongue on the shank or tang of the drill, because it progresses round with the spiral twist as the drill is ground away. Most modern improvements in these devices have incorporated some means of simplifying the setting of the drill so as to speed up operations and minimise the risk of error.

## Tool and Cutter Grinders

In re-sharpening cutters which have a number of edges symmetrically disposed around the axis, perfect control is necessary to preserve this symmetry, and also angular correctness. The grinding machines used for this purpose are, therefore, equipped with longitudinal and cross slides in addition to means of mounting the cutter concentrically to its axis, and indexing it so as to bring the teeth successively into grinding position. Machines of this type are illustrated in Figs. 3 & 11. Their design and equipment varies considerably, but in all cases an important feature is the prevision of universal movements, so that all types of cutters

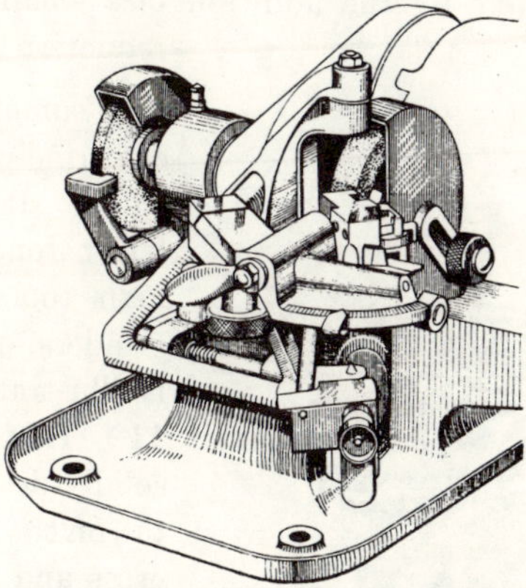

**Fig. 10.—" Perfect Point " twist drill grinder**
(*Jones and Shipman, Ltd.*)

can be effectively dealt with. The scope of the machines is also increased by the provision of various fittings and attachments, and practically all of them are capable of carrying out external and internal cylindrical grinding are surface grinding. A number of examples of cutter grinding are shown, which will explain the adaptability and manner of using these machines.

The "Alba" series of tool and cutter grinders are made with provision either for independent motor or lineshaft drive, in both cases being equipped with a three-speed cone pulley. In the design of the grinding head itself, there is much in common with the plain tool grinder, and one wheel is equipped with a simple form of twist drill grinding jig. The universal tool grinding slide rest is mounted on a knee fitted to a vertical slide on the pedestal of the machine. It is thus capable of a good range of rising and falling adjustment, and the compound slide can be swivelled to any required angle relative to the wheel axis. A removable top slide, similar to a small lathe bed, is fitted with a headstock and a tailstock, both capable of being adjusted to any position along the bed. The headstock may be equipped for power rotation by the addition of a small motor and reduction gear.

A somewhat different principle is applied to the design of the latest Jones and Shipman tool and cutter grinders, one of which is illustrated in Fig. 11. This embodies a solidly-designed pedestal, on the top of which cross and longitudinal slides, similar in design to those on cylindrical

**Fig. 11.—Universal tool and cutter grinder**
*(Jones and Shipman, Ltd.)*

grinders, are fitted. A bed with a universally-swivelling headstock (Fig. 12) and a sliding tailstock is superimposed on the top slide and has a limited swivelling movement about its own centre on the latter. The grinding head is mounted on a hollow cylindrical pillar which slides vertically

**Fig. 12.—Swivelling work-headstock of Jones & Shipman grinder**

in a socket in the pedestal casting, and may be raised or lowered by a handwheel. At the lower end is mounted a bracket carrying the driving motor, which drives the wheel spindle by a belt enclosed in the pillar. The sliding surface of the pillar, where it emerges from the top of the pedestal, is protected from particles of abrasive by a telescopic guard. Swivelling of the head to any angle relatively to the bed is also provided for. The combination of all the sliding and swivelling movements, in conjunction with a number of fittings and attachments to the cutter headstock, makes it possible to adapt the machine to deal with any type of cutter, in addition to cylindrical and surface grinding operations. A power-driven headstock for cylindrical grinding is shown in Fig. 13.

## Indexing Cutters

In order to bring each tooth of a milling cutter into its correct position for grinding, some means of indexing

**Fig. 13.—Motor - driven work headstock of Jones & Shipman grinder**

is necessary. The simplest and most common device for this purpose is a spring detent or "tooth rest," which is attached to an adjustable pillar projecting either from the sliding table or the wheel head, and engaging the teeth of the cutter as shown in Fig. 14. Provided care is taken to eliminate backlash, this device serves the required purpose in the majority of cases, with either straight or spirally-fluted cutters, and in the latter case is the only practicable device of reasonably simple design. Positive division plates are, however, necessary or desirable in some cases, and may be attached to the cutter mandrel, the upper surface of the headstock being adapted to carry a bracket for an indexing plunger (Fig. 15).

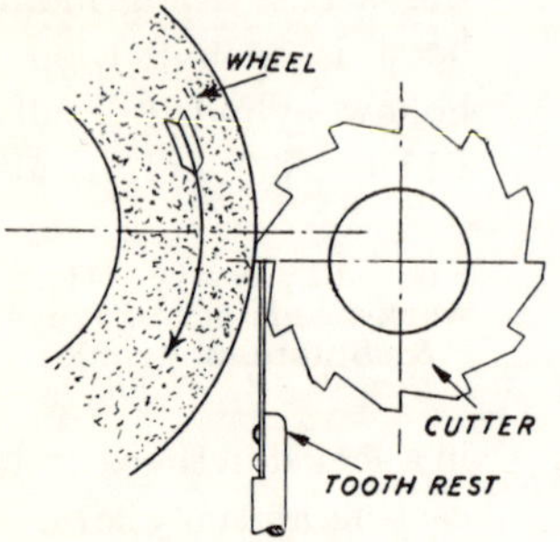


**Fig. 14. Usual method of indexing cutter teeth for grinding**

In grinding an ordinary milling cutter, it is mounted on a mandrel between centres on the work table, and the latter adjusted parallel to the grinding spindle axis. The usual method of presenting the cutter to the wheel is as shown in Fig. 14, where it will be seen that the cutting face of the tooth being ground abuts against the tool rest, and is kept pressed against it by the force exerted by the grinding wheel. The cutter mandrel is adjusted to a lower level than the grinding spindle, in order to produce the clearance angle on the teeth. Each tooth is ground at the same cross setting of the table to ensure that all cutting edges are on the same concentric circle ; in other words, that the cutter runs truly. As each is finished, the table is run longitudinally so that the cutter is clear of the wheel, and the mandrel is turned to bring the next tooth into position. The tooth rest must not be shifted throughout the operation.

The method of mounting the cutter shown in Fig. 16 is

sometimes preferred, as it avoids raising a burr on the edges of the cutter, and also reduces the tendency to draw the temper of the tooth; but as the rotation of the grinding wheel in this case tends to turn the cutter away from the tooth rest, great care is necessary to hold it in position, because if it turns during grinding, the tooth will be ruined.

Where fine and exact clearance angle is necessary, the hollow-ground edge produced by the periphery of the grinding wheel is undesirable, and the use of a cup wheel is preferred (Fig. 17). The cutter mandrel is placed at right-angles to the wheel spindle and the tooth rest adjusted below centre so that the radial line of the cutting edge forms an angle, $x$, with the horizontal; this angle is equal to the clearance angle. The set-up may be inverted if desired, the same precautions then being necessary as in peripheral grinding under similar conditions.

**Fig. 15.—Positive indexing device on Jones & Shipman work headstock**

Figs. 18 to 25 show various applications of tool room grinders, which are self-explanatory, and call for no special comment.

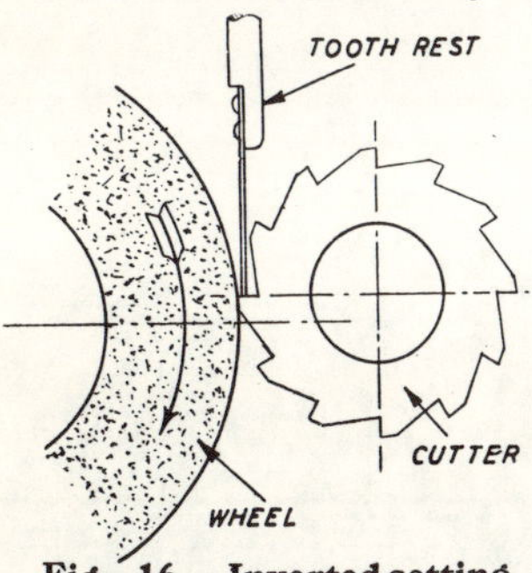


**Fig. 16.—Inverted setting for cutter grinding**

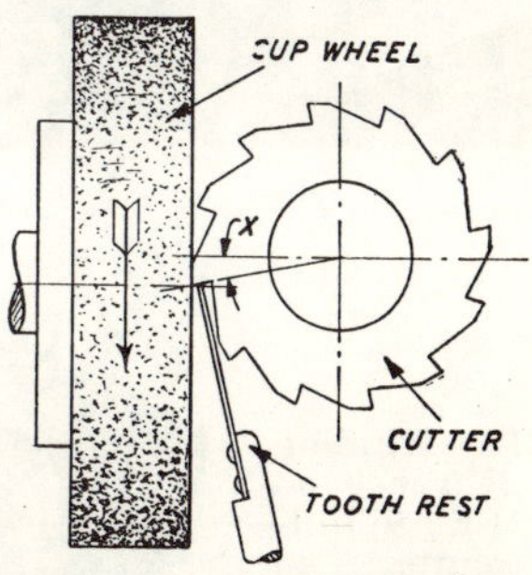


**Fig. 17.—Grinding cutter on side face of wheel**

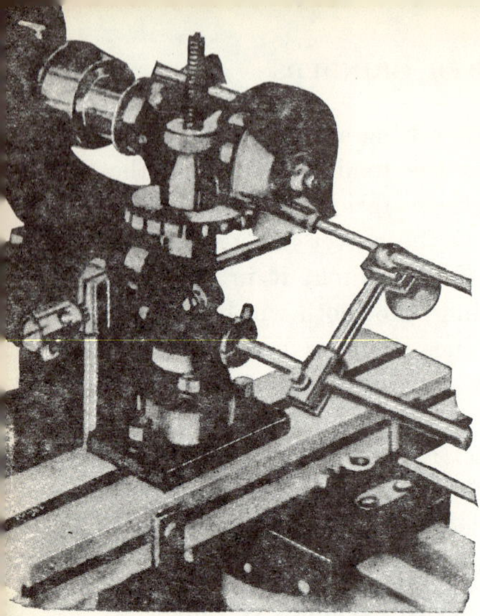

Fig. 18.—Grinding cutting face of teeth of large milling cutter on "Alba" grinder

Fig. 19 (below).—Grinding slab mill with cup wheel on Jones & Shipman grinder

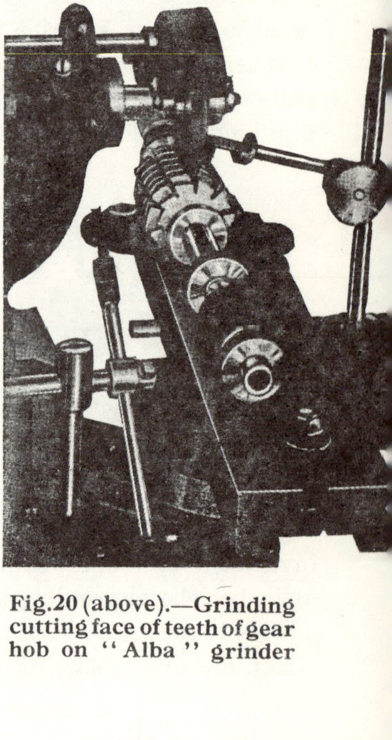

Fig.20 (above).—Grinding cutting face of teeth of gear hob on "Alba" grinder

Fig. 21.—Grinding side clearance on teeth of large milling cutter in Jones & Shipman grinder

Fig. 22.—Cylindrical grinding in 'Alba' grinder

Fig.23 (above).—Grinding machine reamer on Jones & Shipman grinder

Fig. 24 (above).—Grinding inserted-tooth milling cutter with cup wheel on Jones & Shipman grinder

Fig. 25.—Grinding circular saw on "Alba" grinder

## CHAPTER V

# CYLINDRICAL GRINDERS

A PARTICULAR feature of all modern cylindrical grinders is their massive and rigid structural design. Although the torque stresses and the forces tending to push the work and the tool head apart are very much less than in lathes, or other machines using cutting tools, the high precision demanded of grinding machines calls for the utmost steadiness of support, both for the work and the bearings of the grinding wheel.

It has been stated that the very small side pressure necessary to maintain the cut of a grinding wheel tends to facilitate accuracy by eliminating or considerably reducing spring of the machine structures or slides. The comparatively small increments in which the grinding process normally proceeds is also a help in producing accurate work, but it is important to remember that these properties can only be usefully exploited by observing the most scrupulous accuracy in the construction and adjustment of the machine itself. It is a gross fallacy to suppose that a ground finish to any component implies, in itself, any guarantee of precision; grinding is only a precision process when it is done skilfully on a machine constructed and adjusted to the most meticulous limits of accuracy. Bad work can be, and unfortunately often is, produced by grinding, if these simple facts are ignored.

There is no single working component of a grinding machine which can be dismissed as unimportant. The slides, the screws or other gear for operating them, work headstocks and spindles, wheelhead slides and bearings, all

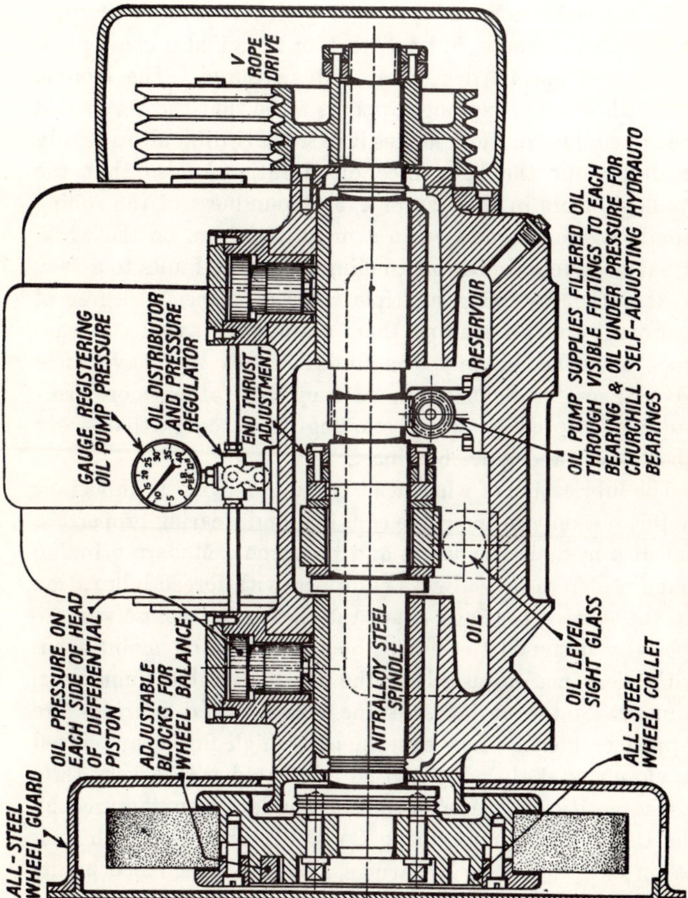


**Fig. 26.—Section of wheelhead of Churchill grinder, showing "Hydrauto" self-adjusting bearings**

have very important and far-reaching effects on the quality of the work produced.

## Wheel Bearings

Many machines have been made with ball or roller bearings to the wheel spindle, but for work of the highest class, plain bearings of special design are still favoured. The trouble with ball or roller bearings appears to be that the very small area of contact in them makes it extremely difficult to steady or damp out the last trace of vibration ; also that the smallest errors in the size or exact roundness of the rolling members are reproduced in a magnified form on the work. It has been found possible to eliminate these faults to a very great extent by using multiple races of a special degree of accuracy, and pre-loading them endwise to avoid the least trace of shake. Spindles mounted in this way have some advantages for light work, as they are rather more free-running (especially when starting from cold) than very closely-adjusted plain bearings.

The lubrication of wheelhead bearings is a most important factor, not only in their free running, and wearing properties, but also in their steadiness and precision. Modern grinding machines are nearly always equipped with forced lubrication to the wheel bearings, and the bearings themselves are specially designed to enable the oil film to be maintained, with clearance adjusted to the finest possible limits. In some cases the surfaces of the spindles and bearings are lapped or honed to an exceptionally high finish, and fitted so closely that they have to be lubricated with a very thin, low-viscosity oil. Many types of grinders manufactured by the Churchill Manufacturing Co., Ltd., are fitted with their patented " Hydrauto " bearings, (Fig. 26), which incorporate automatic adjustment, and are lubricated by a constant flow of filtered oil delivered by a submerged rotary pump. The " Precimax " series of grinding machines, manufactured by Messrs. John Lund, Ltd., incorporate the patent " Oilwedge "

Fig. 27.—Churchill model BY external grinding machine

bearings, in which the oil is distributed by grooves tapering off in depth, which have the effect of building up a wedge-shaped oil film, on which the spindle runs with the minimum friction and the utmost steadiness.

Driving motors built into the wheelhead spindles are used in some cases, but there are several practical disadvantages in this method, and the most popular arrangement at present is by means of a multiple vee-belt drive from a motor mounted at the back of the wheelhead. Very few machines nowadays use the old flat belt countershaft drive.

## Types of Grinding Machines

Some machines are specially designed for external grinding only, and are generally referred to as "plain" grinders; examples of this type, illustrated here, include the Newall Model L (Fig. 2), the Churchill Model BY (Fig. 27), and the "Precimax" Series MPL (Fig. 28). Another class is made for internal grinding only; in this case the work headstock is equipped with a chuck, and no tailstock is fitted. The Churchill internal grinders have the wheelhead mounted on the cross slide of a bridge or gantry over the top of the work slide, as in the case of the Models HBY and A.C. machines illustrated (Figs. 29 and 30). The latter is a special type of machine adapted for very deep grinding jobs such as hydraulic cylinders. The overhang of the work from the chuck is

Fig. 28.—"Precimax" series MPL external grinding machine

supported by a three-point steady, which is adjustable to the required distance from the chuck.

A third type of cylindrical grinder is readily adaptable to external or internal grinding, either by interchange of the wheelhead, or by provision for bringing alternative wheelheads into working position, without actually dismounting either from the machine. By this means, the scope of a single grinding machine is almost doubled, without any great sacrifice of efficiency in comparison with the more highly specialised machines (Fig. 31).

It is, of course, necessary to run the internal grinding spindle very much faster than that for external grinding, and the belt driving gear is thus arranged, either by alternative large and small driving pulleys, or by using a two-stage drive, to provide for changing speed. The Churchill Model PBH grinder, illustrated in Fig. 32, incorporates a countershaft on a swinging arm, mounted over the main wheelhead, and the internal grinding spindle is also mounted on a similar arm so that it can be swung down into working position and rigidly bolted against the front face of the main wheelhead.

## Cylinder Grinding Machines

Cylinder blocks and similar heavy or awkwardly-shaped components which have bores to be accurately machined, are very difficult to set up to rotate on the bore axis, and in many cases this could only be done by using a machine of a size entirely disproportionate to the actual operation. The boring operations on such parts are usually carried out by bolting them to the table of a horizontal boring machine and using a rotating boring head ; similar principles have, therefore, been developed for grinding them. An example of this type of machine is the Churchill No. 1. Cylinder

**Fig.29.—Churchill model HBY internal grinding machine**

**Fig. 30.—Churchill model AC internal grinding machine**

**Fig.31.—"Precimax" Universal grinding machine**

**Fig.32.—Churchill model PBH Universal grinding machine**

THE CHURCHILL
MACHINE TOOL Cº Lᵀᴰ
MANCHESTER
ENGLAND

THE CHURCHILL
MACHINE TOOL Cº Lᵀᴰ
MANCHESTER ENGLAND

Grinder illustrated in Fig. 33. It embodies a headstock mounted on a vertical slide, which is mounted on a longitudinal sliding carriage. At the opposite end of the machine is mounted the work table, which has cross sliding adjustment, and is equipped with fixtures for holding the work. The headstock spindle is geared to rotate at slow speed, and is bored to admit an internal housing which carries the grinding spindle, the drive to which is independent of that of the outer spindle, and provides for running the wheel at the proper grinding speed. This internal housing is not fixed in concentric relation to the outer spindle, but may be adjusted to the required degree of eccentricity to suit the bore of the cylinder which is being ground. This adjustment is so contrived that it can be altered while the machine is running, to feed the wheel into cut.

The work is set up by suitable adjustment of the headstock slide and the work table cross slide, until the bore to be dealt

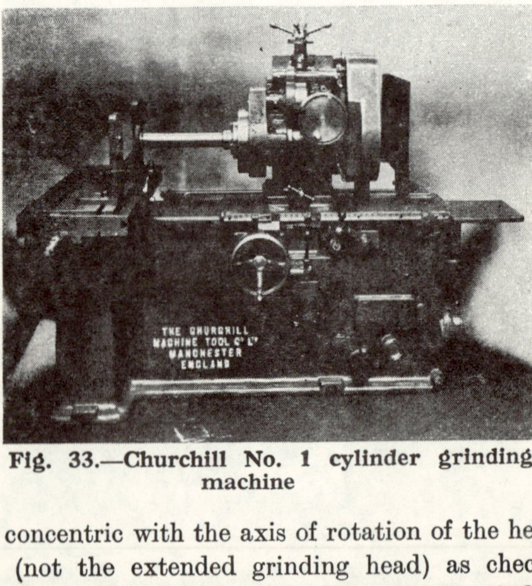


**Fig. 33.—Churchill No. 1 cylinder grinding machine**

with is concentric with the axis of rotation of the headstock spindle (not the extended grinding head) as checked by measurement from the axis of the headstock spindle.

Machines of this type are often termed "planetary" grinders, as the wheel rotates on its own axis and also around the axis of the headstock spindle. One of their special merits is that their method of operation always produces a parallel hole, though not necessarily in a perfectly correct plane; the latter is influenced by the way the work is set up on the work table, and care is necessary to ensure that the bore axis is in alignment with that of the headstock in both horizontal and vertical planes. Where it is convenient, the work is usually mounted by the base flange, or other suitable machined reference face, to a right-angled fixture on the work table. Once the latter has been initially adjusted, so that it is exactly square with the headstock axis, any number of pieces may be subsequently set up without difficulty. In some cases it is also possible to locate the work by a spigot or other positive location surface, so that its concentric adjustment also becomes automatic after the fixture has been properly set up.

### Hydraulic Feeds

The advantages of hydraulic feed to the slides of grinding machines, as compared to mechanical feeds by geared screws or racks, has long been realised, and most modern machines are equipped in this way. Not only does hydraulic feed make possible the simplest and most flexible control at any required rate of travel, but it is smoother in action than mechanical feed. Faults in the work, such as waves, ridges or ripples, often have been traceable to errors or lack of smoothness of the mechanical feed gear.

Hydraulic feed was at first applied only to the longitudinal slides of grinding machines, but is now extensively used for all feeds. The essential mechanism of hydraulic feed gear is intrinsically simple, consisting, as shown in Fig. 34, of a piston which fits inside a double-ended cylinder, with provision for admitting water, oil, or other liquid to either end under pressure. The pressure is generated by a pump, which on modern machines is usually driven by a separate motor. The control valve which serves both ends of the cylinder is so designed that, when one end is open to pressure, the other is open to exhaust, and *vice versa*. By varying either the pressure or the area of the passage by which the liquid enters the cylinder, the rate of feed can be controlled. The piston is coupled to the table of the machine by a flexible linkage so that no force except that in the direction of motion is exerted on the slides. The control valve is operated automatically by dogs on the sliding table, which may be set for any desired length of stroke, and are arranged to throw over the control valve lever to one side or the other as they reach the end of their travel.

Modern practice in cylindrical grinding favours a comparatively slow rotation of the work and a rapid traversing feed, as this distributes the work evenly over the wheel face, instead of chiefly at the sides of the wheel, as happens when very slow traversing feeds are used. The work speed may be from 15 to 60 feet per minute, and traversing rates from about one-quarter of the width of the wheel face per revolution, for light work, to half the width for heavy work. The wheel should always be run at the surface speed recommended by the makers for the particular grit, grade and material being ground. In all cases the work being ground rotates in the same direction as the wheel, so that the surface in contact with the wheel opposes that of the latter ; this sounds contradictory, but a little thought will show that it is correct, as two circles rolling together so that adjacent surfaces move in the same direction will actually *rotate* in opposite directions.

The depth of cut taken in grinding will vary according to the nature of the work, the finish desired, the width of the wheel and the available power. Roughing cuts as heavy as .004 in. are taken on heavy machines, but about .001 in. is more normal. Light cuts are usually from .0005 in. down to .00025 in. In general, wet grinding allows heavier cuts than dry grinding, because not only is the heat effectively conducted away, but the cutting action is entirely free from the clogging effect of waste particles, as they are immediately washed away. When finishing off a surface, it is usual to allow a dwell after feeding the wheel to the required depth, so that the effect of any possible spring is eliminated. This is sometimes referred to as "sparking out," because in dry grinding operations, the traversing motions are continued

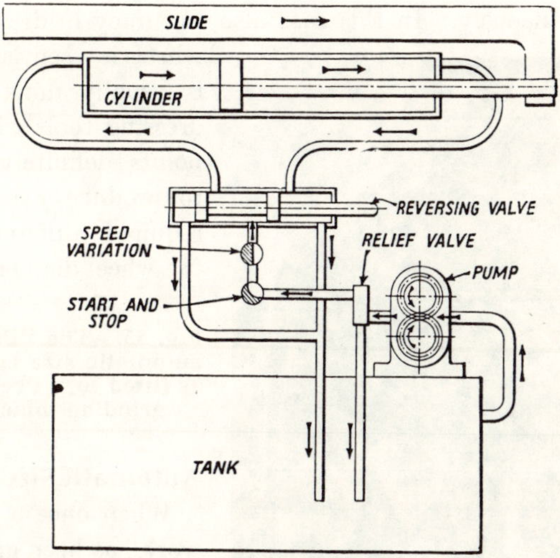


Fig. 34.—Layout of hydraulic feed system of Churchill grinder

after the in-feed is concluded, until the wheel ceases to produce sparks from the work.

## Plunge-cut Grinding

Operations can be speeded up by using a wide wheel and a short traverse, provided the power available is adequate, and that the wheel face is dead true with the direction of traverse. This practice is becoming very popular in modern production work, and machines are equipped for automatically feeding the work in at the required rate for roughing, after which the feed tapers off to the final dwell for finishing. Wheels up to 8 in. wide are used on many modern machines.

Apart from the power requirements, and the necessity for very rigid heads, plunge-cut grinding demands perfect truth of the wheel face, obtained by the use of a dressing tool fitted to the work slide. No wheel, whatever its width, can produce its best work unless it is kept perfectly true on the face.

Machines are sometimes fitted with two wheels of different diameters, to grind two different-sized parts of a shaft simultaneously. In this case also, accuracy in dressing the wheels is essential, and it is usually done by two dressing tools with their points definitely located to produce exactly the required differences in the wheel diameters.

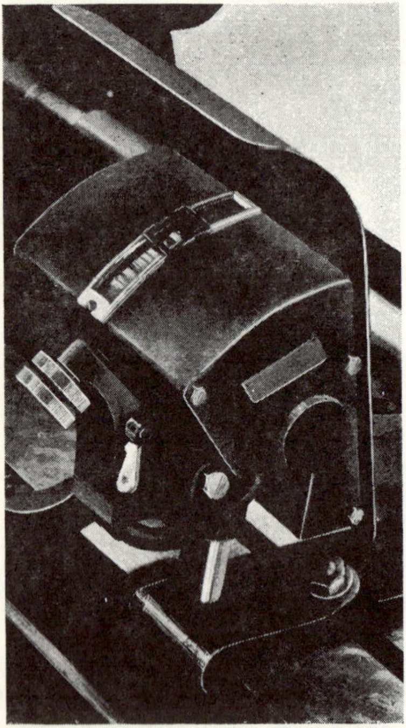

**Fig. 35.—The "Preciser" automatic size controller as fitted to "Precimax" grinding machines**

### Automatic Size Control

When once a piece of work has been ground to the correct size, within fine limits, it is possible to set the controls of the machine so that the infeed of the table is positively stopped at the same point each time; thus it is logical to assume that subsequent pieces made to this setting should also be the correct size. It is a fact that in a rigid and properly adjusted machine, a large number of pieces can be produced to fine limits in this way. The provision of automatically graded in-feed, with final dwell, assists in facility in maintaining uniformity by eliminating spring; but sooner or later, the effect of wear on the wheel will result in producing a slight difference in diameter. Exact control of size can, however, be obtained by means of automatic gauging devices which actually measure the work while it is in progress. These may be used to assist the operator by advising him when the correct size has been obtained, but in their latest development it is also possible to use them actually to control the machine. The "Preciser" Automatic Control (Fig. 35) used on the "Precimax" grinders of Messrs. John Lund, Ltd., incorporates an automatic gauging device which, as soon as the required size has been reached, energises a photo-electric relay. The current from this, suitably amplified, controls mechanism which stops the feed and returns the table to the unloading position.

**Fig. 36.—Swivelling motor-driven headstock of "Precimax" Universal grinder**

## Work Table Design

A special feature of all modern grinding machines is the complete protection afforded to all sliding bearings, by designing the upper slides so that they overlap the latter in all positions of travel. The upper work table is in most cases made in the form of a large tray, having a flat and usually raised surface, on which the bed carrying the work centres is mounted, with adjustment for taper turning by swivelling the entire bed. Owing to the length of the latter, extremely fine adjustment of taper is possible by this method, and setting is further facilitated on many machines by the use of a micrometer screw. In addition to this movement, it is usually possible on universal grinding machines to swivel the work headstock, in some cases through a full 90 degrees, for dealing with steep tapers on work held in the chuck (Fig. 36). If the wheelhead is also fitted to a short swivelling slide, it is possible to grind two different tapers at one setting of the machine.

Work headstocks are nearly always driven by a separate motor on modern machines, the necessary speed reduction being obtained by gearing or compound vee-velt. Variable-speed motors are often used in order to obtain a range of work speed to suit jobs of different diameters.

## Steadies

When long, slender shafts have to be ground, there is a tendency for them to sag in the middle, and although the machine may be accurately set for parallel grinding, the work will be larger in the centre than at the two ends. In order to prevent this, one or more steadies may be rigged on the bed to support the back and underside of the work. In the case of very long shafts, as many as half a dozen steadies may be found necessary to ensure perfect parallelism. (See Fig. 32).

In most cases, the steadies, or "back-rests," as they are sometimes called, have independent adjustments in both horizontal and vertical directions, but sometimes a very simple bevelled steady with unilateral adjustment is used. Steadies of any kind should never be used on a shaft that is running out of truth, as possibly it may be after hardening. The shaft should be made as straight as possible and then skimmed up to exact concentric truth by light cuts with the wheel, before the steady is rigged.

## Centreless Grinders

Simple and straightforward grinding operations, including finishing and sizing such parts as pivot pins, rollers, long rods, and even plungers and pistons, may be carried out expeditiously on centreless grinding machines. In these machines, an example of which is seen in Fig. 33, no work-carrying headstock is provided, but the work is supported on a flat or bevelled rest running parallel to the face of the grinding wheel, and rotated by the action of a feed wheel

**Fig. 37.—" Record " centreless grinding machine**
*(B. Elliott and Co. Ltd.)*

on the opposite side. This runs much more slowly than the first wheel, and drives the work at the required peripheral speed by friction. The feed wheel head is mounted on a slide so that it may be advanced towards or retracted from the grinding wheel to adjust the size of the work, and it may also be set at an angle to the latter for taper turning. In order to feed the work laterally across the face of the grinding wheel, the feed wheel may be inclined in the vertical plane, so that it produces a spiral turning effect on the work and thus moves it endwise at a rate determined by the angle of inclination.

Parallel parts may be fed once across the wheel face at such a rate that they emerge finished to size on the other side. Work having a shoulder or a taper may be fed through to a limited extent, as determined by a stop rod (which may also act as an ejector) coming into operation either manually or automatically at the completion of the stroke ; the feed wheel is simultaneously withdrawn from the grinding wheel to prevent further grinding action. Machines of this type may readily be adapted for use with a magazine feed, so that they become practically automatic in operation, and speed up production to the utmost.

## Thread Grinding

Screw threads in which specially high accuracy and finish are required are also produced or finished by grinding methods, using machines which incorporate principles comparable to those of the screw-cutting lathe. The grinding wheel may be dressed to a single vee-point of correct form for the thread to be ground, equivalent to a single-point screw-cutting tool, or to a serrated edge, equivalent to a machine chaser. While the former is simpler to keep dressed to the correct shape, the latter has the advantage of having a much greater wearing surface, and thus requires much less frequent dressing. Thread grinding machines are often constructed so that the work spindle rotates in alternate directions, and the wheel is kept in continuous action both ways. In this case, the utmost care is necessary to take up backlash in the gearing and lead screw.

The wheel spindles of thread grinding machines must be capable of adjustment to an angle corresponding to the pitch angle of the thread. Wheel speeds are higher than usual, from 7,500 to 8,500 f.p.m., and a hard vitrified bond is preferred, from 120 to 150 grit or finer; both speed and grading are critical if accurate work is to be produced and wheel form maintained.

Among the common applications of thread grinding are the finishing of high-grade screw taps, screw gauges, etc., after hardening, also hardened worms for transmission gearing, an example of the latter being shown in Fig. 38.

## Cam Grinding

The machines used for cam grinding incorporate means of rocking or reciprocating the work centres to and from the grinding wheel (or *vice versa*), this motion being controlled by means of a master cam, which rotates at the same speed as the work. A number of cams on a single shaft may be dealt with on a production machine which selects and indexes each cam in turn by automatic mechanism. Machines used for piston grinding are sometimes arranged to give a slight reciprocating motion to the work so that it is ground very slightly oval, the object in this case being to provide increased working clearance in the plane of the gudgeon pin.

**Fig. 38.—Grinding thread of hardened transmission worm**

*(Alfred Herbert, Ltd.)*

# CHAPTER VI

## SURFACE GRINDERS

THE application of grinding processes to the production of flat surfaces has developed rather slower than cylindrical grinding processes, but is now just as fully established as an essential department of engineering practice Many of the requirements and features of design of surface grinders are identical to those of cylindrical grinders, or at least differ only in so far as is obviously necessary to adapt them to a different order of motion As already indicated, surface grinders broadly resemble milling machines in their motions and the nature of the work they deal with ; and like milling machines, generally have the spindle disposed either horizontally or vertically.

The relative merits of horizontal and vertical machines can only be assessed with reference to the conditions of use, and the nature of the work handled, in individual cases. In tool-room work, horizontal machines are very popular, because they have a very wide scope, and are particularly adaptable to the everyday kind of tool-room jobs, such as grinding jig plates, parallel gauge or packing slips, etc. Similar operations which occur in production practice are equally well handled by horizontal grinders.

For heavy grinding on continuous flat surfaces, or finishing of large areas, however, the vertical grinder is generally preferred. Not only is it usually more efficient in the rapid removal of material, because of the large area of grinding surface which can be brought into action, but finish and accuracy are also facilitated by taking a cut equal in breadth to the diameter of the wheel; on the other hand, any inaccuracy in the relative squareness of the table and the wheel spindle, or faulty design or adjustment of the bearings, would have more serious effects than in a horizontal spindle machine. The latter are also necessary for dealing with work having grooves, or strips at varying levels, to be ground, as the vertical spindle machines, broadly speaking, are only suitable for plain flat surfaces.

## Horizontal Grinders

Modern machines of this type have many features of design in common with cylindrical grinders, including automatic feeds for traversing and cross movements; in many cases these feeds are hydraulically operated. Vertical feed is usually provided by mounting the wheel-head on a vertical slide, equipped with micrometer feed, and in cases where the wheel is belt-driven, the necessity arises for some form of compensating gear to maintain the belt at correct tension, irrespective of the height of the wheel. The "Pallas" grinder (Fig. 39) incorporates a system of jockey pulleys enclosed in the pedestal of the machine, the

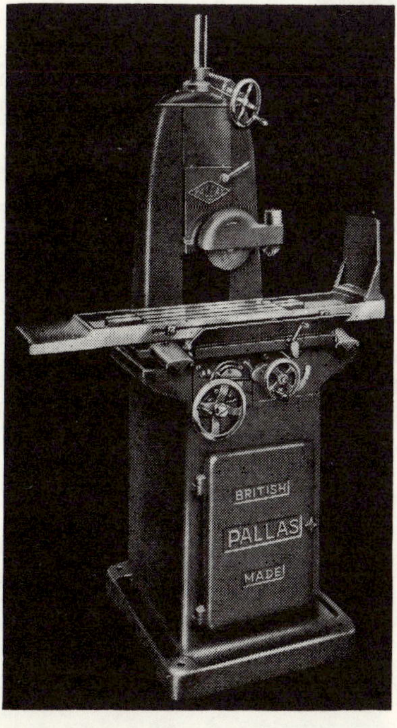


**Fig. 39.—"Pallas" horizontal surface grinder**
*(B. Elliott and Co. Ltd.)*

Fig. 40.—Churchill model NB horizontal surface grinder

driving motor also being enclosed. In the Churchill model NB machine (Fig. 40), a long box slide is employed, having the wheelhead attached to the top and the driving motor to the bottom of it. The driving belt is outside the box slide, at the rear of the machine, and is enclosed by a detachable cover. A somewhat similar arrangement is employed on the Jones & Shipman machine (Fig. 41), but in this case the motor is attached to the side of the column and the belt drive is taken obliquely to the wheelhead.

The Mattison surface grinders, which are made in a range of sizes for traverses up to 12 ft., have the driving motor built into the wheelhead, and on the larger machines the vertical feed is power-operated. A special feature of these machines is the double-column vertical slide, which ensures the maximum rigidity, and permits of ready adjustment to take up wear ; and also the cross slide which supports the motor and wheelhead, so that the work table is only required to make longitudinal movements.

## Wheel Bearings

The requirements of these are much the same as for cylindrical grinders, and if work of the highest accuracy is to be carried out, are just as exacting. Plain bearings, in some cases supplemented by ball thrust races, are generally favoured, and special provision for forced lubrication is often incorporated. The Churchill grinders employ the patent self-adjusting "Hydrauto" bearings, as described in the preceding chapter.

Wheel dressing is carried out by means of a diamond tool mounted vertically in the socket of a fixture which may be clamped to the work table and fed across the wheel face by the cross feed. In order to ensure that the diamond is presented at the correct angle to the wheel, it must be located exactly under the centre of the spindle, and in order to facilitate this setting, the Churchill Model NB Grinder is equipped with a vertical gauge rod, fitted to a socket on the front of the wheel casing, and capable of being brought down over the diamond tool to check its alignment.

As in the case of cylindrical grinders, it is important to protect all the sliding bearings of these machines from abrasive dust, and this applies in particular to the vertical slides which carry the wheel-head. In many cases, cover strips or casings are fitted which enclose the actual slide surfaces. The table slides are similarly protected, and screw-operated slide motions, are so designed that the threads cannot become exposed to the action of the dust at any position of the slide travel.

Although most horizontal

Fig. 41.—Jones & Shipman horizontal surface grinder

surface grinders are intended exclusively for working on horizontal surfaces, using the periphery of a plain disc wheel, the possibility of grinding vertical surfaces, with an annular or " cup " wheel, is not definitely excluded. Some machines are designed particularly for grinding the edges of slabs or angle plates in this way, and in such cases segmental wheels similar to those used on heavy vertical grinders are commonly employed (Fig. 42).

The principle of the planing machine is sometimes adapted to grinding, by substituting one or more wheelheads for the tool slides ; the rest of the machine follows normal practice, except that the power feed to the table needs much less power and moves both ways at equal speeds. Machines of

**Fig. 42.—Lumsden horizontal grinder with segmental wheel for edge grinding**

this type are often used for finishing the beds of machine tools, including raised and side vees in addition to flat surfaces ; these are all ground simultaneously by multiple wheelheads set to appropriate angles.

## Vertical Grinders

The vertical slide and wheelhead structures of these machines require to be extremely robust in design because of the weight which must be rigidly supported, and the spindle bearings must incorporate provision for supporting heavy end thrust, both upwards and downwards. Driving motors are in the majority of cases built into the wheelhead, avoiding the need for belt drive, and, as previously mentioned, considerable power is necessary in view of the large wheel area in active operation. The wheels are, of course, designed to work on the underside face, and may be of the solid annular type, but usually, and especially on the larger machines, segmental wheels are employed. These consist of metal wheels having cavities formed in the rim, into which segments of abrasive material are secured, by means of wedges and screws, so as to enable them to be moved forward or completely replaced when worn. This not only results in an economy of abrasive material, but also improves the cutting action, due to the relief afforded by the slots between the segments. The wheels have to be carefully balanced, and the cutting face is dressed to perfect flatness in the same way as a solid wheel.

The longitudinal table feed is usually operated by hydraulic gear, and in many cases no cross feed is provided, the spindle being directly over the centre of the table width, and the wheel of sufficient diameter to span it completely, so that any work mounted on the table will be ground as it passes under the wheel. Vertical feed may be provided by the slide which carries the wheelhead, but in some cases the latter is fixed and the spindle adjusted endwise by a telescopic ram. In some machines the complete wheelhead is capable of being moved vertically for coarse adjustments, such as for accomodating work of various heights, but is clamped under working conditions, and fine feed adjustment obtained by the spindle traversing gear.

Some types of vertical grinders have been made with the spindle axis permanently or adjustably set at a very slight angle to the vertical so that only one side of the wheel comes into contact with the work, the other side having a small clearance. This allows of taking deep roughing cuts, and avoids clogging the wheel when mill scale or other foreign matter on the surface of the work is encountered. The work is, however, left slightly concave after grinding.

The " Precimax " vertical surface grinders incorporate a reversible electric motor for raising and lowering the wheelhead, and a hand feed for fine adjustment. The reciprocating table types of " Precimax " grinders are made with tables 12 in., 18 in. and 24 in. wide, with a length of 48 in., 60 in., and 72 in. respectively, but extra lengths of table are also available.

The Churchill vertical grinders, an example of which is seen in Fig. 43, also incorporate a vertical slide adjustment for the wheelhead, which has a built-in motor, and the micrometer feed can be operated either by hand or automatically.

Rotary tables are often employed on vertical grinders, in some cases being capable of continuous operation on work pieces which are secured to the slowly moving table on one side and removed on the other. The wide range of Lumsden grinders includes several such machines, and also a twin rotary table machine, as seen in Fig. 44, in which the wheelhead is mounted on a radial arm which can be swung over either of the tables as required. One table is thus available for unloading

**Fig. 43.—Churchill vertical-spindle surface grinder**

and loading while the grinding wheel is operating on the other.

Horizontal spindle machines are sometimes equipped with rotary tables, but their application in rapid production work is by no means so widespread as vertical spindle machines, for the reasons already given.

## Coolant Systems

Most heavy surface grinders are equipped with a pump and circulating system for cooling fluid. As the wheel has a tendency to scatter the fluid in all directions, it becomes necessary to enclose the work table fairly effectively by means of easily-rigged sheet metal screens. Rotary table machines may have the table and wheelhead almost totally enclosed. The fluid employed in most cases is similar to that used as a cutting fluid on other machine tools, but mixed with considerably more water than usual to form a very thin emulsion.

In tool room work, surface grinding is often carried out dry, one reason for this being that it is undesirable to interfere with visibility of the work any more than is absolutely necessary. A good deal of dust and sparks are thrown off the

**Fig. 44.—Lumsden twin rotary table vertical-spindle surface grinder**

wheel in this case, and in addition to fitting guards to the work table to prevent too widespread distribution of grit, it is now common to fit vacuum extractors to the machines. The nozzles of these are located close to the wheel to intercept as much of the dust as possible, and may be seen in the photos of the machines seen in Figs. 39 and 41.

## Methods of Holding the Work

The work tables of surface grinders may be equipped with machine vices or other fixtures for clamping work securely during the grinding operation. Manufacturing production usually calls for the use of special jigs for this purpose, and in the case of fairly small parts which are ground in batches or "gangs," multiple jigs are often employed. One of the most useful devices for holding iron and steel components is the magnetic chuck, which is particularly adapted to the

**Fig. 45.—"Eclipse" permanent magnet chuck holding work on table of surface grinder, showing use of special packings**

purposes of grinding and is regarded as almost indispensable for many classes of work. The majority of these devices employ electro-magnets, and are thus dependent upon a source of direct current supply for their operation, but in

**Fig. 46.—" Eclipse " permanent magnet chuck holding work for cylindrical grinding**

recent years improvements in magnet steels have made it possible to employ permanent magnets for this purpose.

The range of " Eclipse " chucks, manufactured by Messrs. J. Neill & Co. Ltd., Sheffield, covers all requirements, both for holding work flat on the work table of a surface grinder, or for chucking it on the spindle of a cylindrical grinder. Examples of both these applications are seen in Figs. 45 and 46. As these chucks are entirely independent of electric current supply, and do not involve the use of cables, which might be liable to foul other parts of the machine and become damaged, their use is practically unrestricted, except in the case of non-ferrous and non-magnetic metals. Chucks of this type can also be mounted on sine tables and used to hold work at all sorts of angles, which would involve considerable difficulty with other kinds of clamping devices.

Work which has been held on magnetic chucks may retain residual magnetism, which is not desirable ; the parts are therefore demagnetised by placing them in an alternating current magnetic field for a second or two after the grinding process is finished, and either removing them gradually from the field, or diminishing its effect by means of a variable rheostat.

**Fig. 47.—Wright 8 in. table surface grinder**
*(Wright Electric Motors, Ltd.)*

## Simple Surface Grinders

In addition to the above machines, which may be described as true surface grinders, a number of much simpler machines are employed for producing more or less accurately flat surfaces on work which is not of a very exacting nature. Some of these machines are really on the "border line" between grinding and polishing processes, and therefore do not call for detailed description here, but others are definitely classified as grinding machines, and are very extensively used in engineering practice.

The characteristic of these machines is that the flatness of the surface of the work is produced or controlled, not by machine slides, but by the inherent flatness of either the machine table or the surface of the grinding wheel itself. One of the most popular of such machines is the "table" surface grinder, an example of which is shown in Figs. 47 and 48. It comprises a vertical-spindle motor-driven annular grinding wheel, projecting through a hole in the centre of the work table, which is provided with elevating adjustment. A vertical fence cuts across a chord of the aperture, and also serves to support a disc which fills up the annulus of the wheel, forming a continuation of the table surface. The wheel disc is equipped with a fan which serves to remove the dust produced by grinding.

An ordinary grinding head can be equipped with a work table which enables it to be used for surface grinding, as

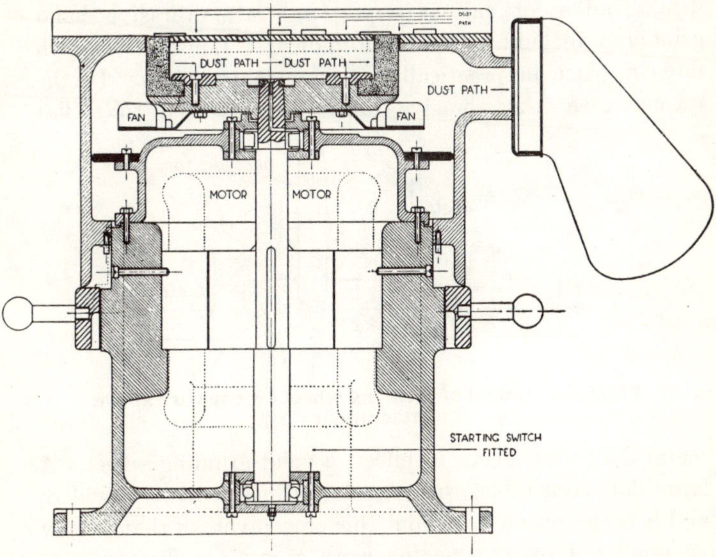


**Fig. 48.—Section of Wright surface grinder**

shown in Fig. 49. The table is provided with a very simple form of elevating adjustment, and the wheel may be trued by means of a dressing tool fitted to a bridge piece, which is

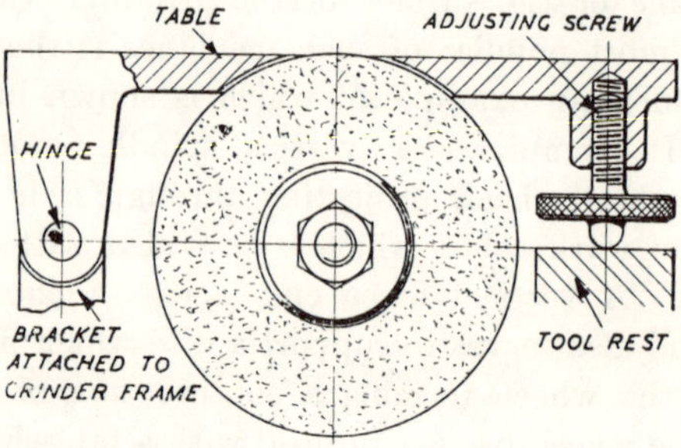


**Fig. 49.—Surface grinding attachment for plain grinder**

guided across the wheel face by a fence clamped squarely across the work table, as shown in Fig. 50.

Abrasive fabric is used in many simple grinders, including disc grinders of various types, and also abrasive band grinders, including the well-known " Linisher " which finds a place in practically every machine shop. In the former case, the abrasive fabric is cemented to a flat

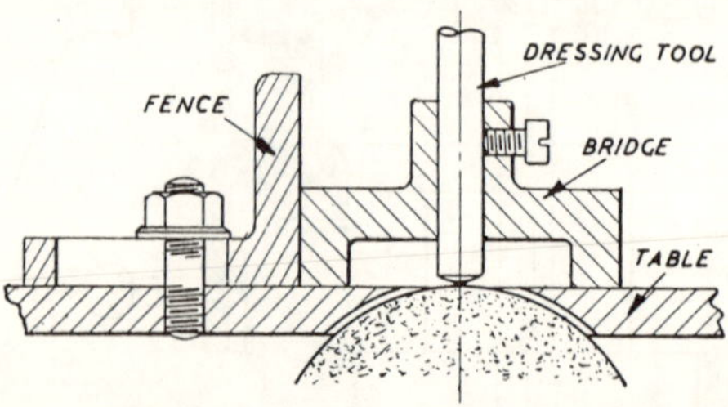


**Fig. 50.—Method of dressing wheel for use with above attachment**

metal disc, and forms, in effect, a solid grinding wheel. A large flat work table, with angular adjustment, is provided, and it is customary to mount this on a pivot, so that it may be oscillated to traverse the work across the disc face, to distribute the wear evenly over its surface.

Abrasive band machines, on the other hand, use flexible bands of abrasive fabric running over two broad pulleys like a machine driving belt (a third pulley is generally

**Fig. 51.—Small abrasive band surface grinder**

employed to tension the band), and a flat table is provided under the top run of the band to support the pressure of the work ground on it. (Fig. 51). Certain types of disc or band machines are very extensively used for finishing woodwork, the process being generally termed " sanding," as the abrasive medium employed is similar in effect, if not identical in nature, to sandpaper or glasspaper.

## CHAPTER VII

# LAPPING AND HONING

THESE two methods have been grouped together in this chapter because they are often used in engineering practice to attain the same ends, and to some extent are applied by similar methods and with similar appliances, but it should be clearly understood that they are separate and distinctive processes. It is necessary to call attention to this fact because there has been a great deal of confusion, even in technical circles, over the precise definition of these terms. Briefly, however, it may be said that both are fine abrasive processes, but whereas lapping is carried out with loose or imperfectly bonded grains, honing entails the use of solid or "fixed bond" abrasive elements, comparable in composition with those used in ordinary grinding.

Lapping has been defined recently by an international authority as "a fine grinding process for producing plane, cylindrical or spherical workpieces of the highest quality, as regards shape and surface, when a loose abrasive and a lubricant are used. The workpieces and the lap are guided in relation to one another, so that they perform cycloidal motions, which result in the equalising or forming of the surfaces." (A. J. Schroeder, D.Sc., *Machinery*, June 27th 1940). The same writer specifies two distinct kinds of lapping processes: (*a*) "equalising" lapping, in which the work and the lap mutually improve their shape and surface during the progress of the operation; and (*b*) "forming" lapping, in which the shape of the lap is imparted to the work.

## Principles of Lapping Processes

The wear which takes place in the working parts of machines, such as in shafts and their bearings, and sliding parts such as pistons and crossheads, is mainly due to the abrasive action of minute particles of solid matter, such as dust, carbon or metal, which are deposited on the surfaces from the atmosphere or carried by the lubricating oil. This fact, if indeed it is not self-apparent, is proved by the reduced wear of mechanisms which are totally enclosed and lubricated with filtered oil, compared with those in which less scrupulous care is taken to exclude foreign matter.

If particles of abrasive material are deliberately introduced between two rubbing surfaces, the " wear " is intensified to a considerable extent, and if suitable measures are taken to control its action, this process may be used for improving finish and accuracy of the surfaces, so long as the abrasive material is completely removed after its mission has been fulfilled. The abrasive action is very much more positive if the cutting particles are mixed with a viscous lubricant such as grease or thick oil, which tends to retard their free movement relative to the surfaces. This process is largely used for "lapping together" working parts, and, in particular mushroom valves of pumps, compressors and internal combustion engines, being popularly, but not quite correctly, known as " grinding in."

This method of lapping together is extremely useful for eliminating or reducing surface inaccuracies (" high spots ") of parts which have to run together—particularly if these surfaces are of such a shape that they cannot readily be finished by available geometric grinding processes. An example of this kind of operation is in the lapping together of gear wheels of all types. This nearly always results in improved smoothness and quietness of running, but is, of course, especially advantageous when the finish of the gears, as originally cut, is known to be imperfect.

Lapping together of mating parts, however, is not always

the best way of finishing them, from the point of view of accuracy, because both the parts wear, at rates which are not necessarily uniform, depending upon the nature and hardness of the respective materials, and the local pressure and rubbing speed. Suppose, for instance that two parts were being lapped together, one being relatively inaccurate, but less susceptible to the action of the abrasive than its more accurate mating part. It is clear that in this case the first part would tend to impose much of its inaccuracy on the second part, and the last state of the surfaces might become even worse than their first. Again, take two discs which are lapped together by concentric rotary motion : the outer edges of the discs will move at a much greater speed than their centres, and thus the abrasive action will not be even ; in short, the discs will not be flat, but slightly convex. This can always be corrected in a simple job of this nature by eccentric motion of the parts, but that is not always possible in work which may have concentric locating surfaces (spigots or recesses), or in the case of mated cones, for example.

Lapping a shaft into a bush, or a piston into a cylinder, may result in producing too great a clearance between the parts because of the amount of room taken up by the abrasive compound. In some cases, attempts to lap together closely fitted parts of this kind may result in seizure, because there is insufficient room to introduce the abrasive properly. For these, and many other reasons, it is often advisable to use separate laps for mating parts in preference to lapping them in contact with each other.

## "Charged" Laps

It is often observed that in the case of two dissimilar metals running together, such as a steel shaft in a babbitt metal bearing, the greater amount of wear takes place, not in the softer metal as might be expected, but in the harder metal. This apparent paradox has been noted in very many kinds of machinery; in clocks, for instance, the brass gear wheels are usually in better condition after many years' running than the hardened steel pinions and pallets. This is because foreign matter which gets between the surfaces becomes imbedded in the softer material, and having no further relative motion thereto, cannot have any abrasive action upon it; the harder material, however, offers no lodgment for the foreign matter, but must continually rub against it.

This principle is utilised in lapping processes using "charged" laps; that is to say, prepared lapping surfaces of material sufficiently soft to enable the abrasive particles to become imbedded in it. In this way the surface becomes practically the equivalent of a grinding wheel, the metal lap forming the "bond," in which the abrasive crystals are held stationary so as to operate effectively on the work to be lapped. The metals commonly used for laps include lead, copper, aluminium and porous cast-iron; also several non-metallic substances such as wood, leather, rubber, and vulcanite or synthetic compositions.

Laps of this kind, if charged with an efficient and suitable abrasive, work with reasonable rapidity, and retain their accuracy quite well, if care is taken to distribute such wear as does inevitably occur fairly evenly over their entire surface. For instance, when lapping parts on a flat lap, or conversely,

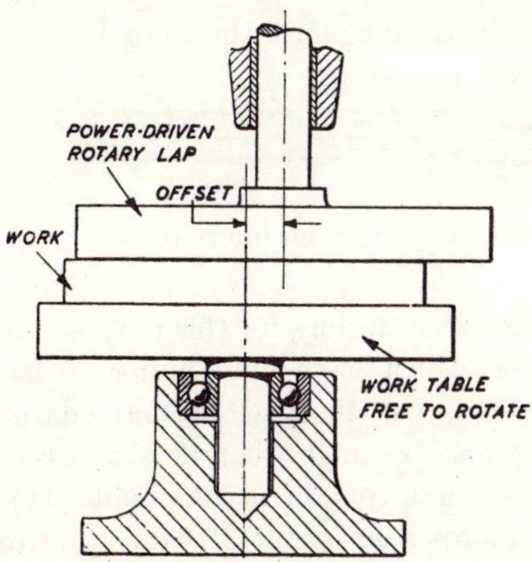


**Fig. 52.—Machine for lapping flat surfaces**

applying the lap to a stationary surface, a form of motion which embraces as much of the area of the lap as possible should be used. Experienced workers often favour a "figure of eight" motion, and partly rotate the work or the lap between strokes.

Machines for flat lapping are generally made, as shown in Fig. 52, with a circular work table, free to rotate on a central pivot, and a rotating circular lap which can be brought down over and dead parallel with it, but eccentric to it to a variable extent. Rotation of the lap also rotates the work table by friction, the relative motion of the two members providing the rubbing friction being due to the eccentricity of the parts. The greater the eccentricity, the faster is the lapping action. In addition to the lapping of flat surfaces, these machines are also capable of lapping cylindrical parts, provided these are so mounted as to be capable of rolling freely between the upper and lower discs.

Laps for internal cylindrical work may consist simply of short plugs of suitable metal, fitted to shanks by means of which they may be rotated, either by hand or machine power. The usefulness of such laps is, however, considerably increased if they are made capable of adjustment to accommodate slight differences in the bore of the hole, or for wear of the lap itself. Spring laps, which press against the bore by their

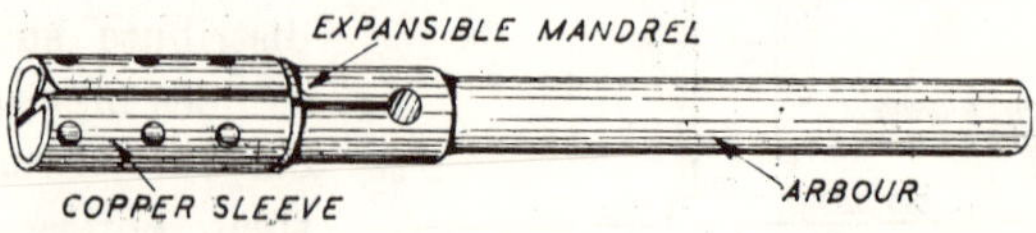


**Fig. 53.—Boyar-Schultz type of internal lap**
*(T. P. Headland, Ltd.)*

own elasticity, or are fitted with springs for this purpose, are useful for finishing bores which have only minor surface irregularities or roughness, but their self-accommodating properties are actually a disadvantage when it is necessary to correct errors of parallelism or roundness. A popular type of lap which is made in sizes from about 3/16 in. upwards, consists of a steel rod, split at one end and equipped with a tapered expanding screw, and having a strip of soft sheet copper, perforated with holes, wrapped around and keyed to it (Fig. 53). The holes in the copper lap retain the abrasive paste, and the sleeve is easily renewed when worn, or changed for working with different abrasives. Internal laps of all types have many practical uses for dealing with small holes which would be difficult to grind internally, or when suitable grinding equipment is not available; they are not, however, very efficient production tools when an appreciable amount of metal is to be removed. Lapping allowance should rarely be more than about .001 in. and may with advantage be very much less.

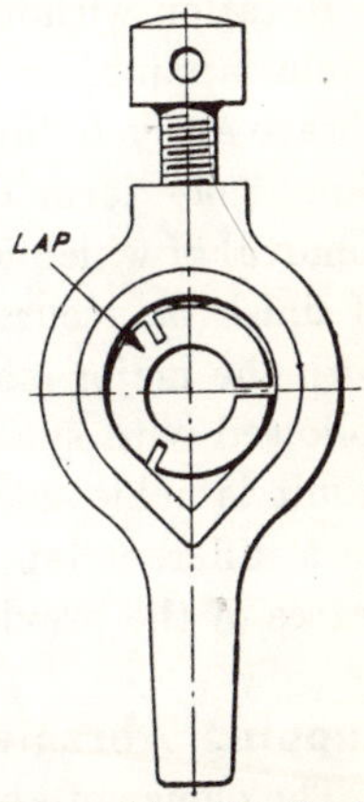

**Fig. 54.—Method of holding split external lap**

External laps may consist of rings or bushes, made of, or lined with, suitable metal to form the lapping surfaces, and preferably with provision for adjustment, such as by splitting and providing some means of contracting, in much the same way as is done with screwing dies. A very simple and effective form of lap consists of a fairly thick bush of copper, aluminium or cast-iron, about the same length as the bore diameter, split right through in one place, and partly through in two or three others; this may be contracted by a band similar to an ordinary hose-clip, or just held in a lathe carrier (Fig. 54).

Either internal or external laps must be applied to the work with a combined rotary and reciprocating movement. It is not really important whether the work or the lap rotates, though individual exponents of the process on particular classes of work often exhibit decided preferences. Special conditions, of course, may emphasise the convenience or efficiency of one method over the other, and a great deal depends on the method or machine used for applying the lap. Drilling machines and lathes are often used for lapping, but in any case the proper control of the process is much facilitated if one or the other member is held in the hand, so that the lapping friction can be felt, and its action thus estimated.

Rotation without endwise motion should never be allowed while lapping, or *vice versa*, or the individual particles of abrasive, by following the same tracks for any appreciable time, may form deep scratches or scores, the subsequent removal of which is difficult. In lapping to a specified degree of finish or accuracy, an abrasive sufficiently coarse to deal with the major inaccuracy or roughness should first be used, followed with successively finer abrasives until the required result is achieved. Each grade of abrasive should be used on a different lap, and its use should be continued until all traces of the previous process are removed.

### Lapping Abrasives

The choice of abrasives for various purposes, and different materials, is extremely important. A complete range of abrasive powders and pastes for all purposes, including carborundum and Aloxite in all grit sizes, is supplied by the Carborundum Co. Ltd., and their advice as to the particular grade which should be used for a particular class of work should be consulted whenever there is any doubt. The vehicle used in making abrasive pastes is also very important, and its viscosity or tenacity must always be proportional to the size of the grains of abrasive. An oil that is too thin to suspend a coarse grit of abrasive properly might be too heavy for a very fine grit.

The hardness of the abrasive also affects its suitability for various purposes. It is essential that the abrasive should always be harder than the material on which it is used, but too hard an abrasive used on a soft metal may result in too fierce and uncontrollable action, and a tendency to become permanently imbedded in the surface. It is, of course, most important that every trace of the abrasive be completely removed when its work is finished, or the surface becomes itself a lap, and sets up destructive wear under normal working conditions. Some abrasives are far more difficult to remove than others; this applies particularly to those with very acute-angled crystals which readily penetrate the surface of soft metals or into the pores of harder metals. The most scrupulous cleaning is always necessary after lapping. Special abrasives, having crystals which can be broken down chemically, or by solvents, so that their cutting ability can be destroyed, have been used for some purposes.

Many natural abrasives, such as the silicates and aluminates —sand, brickdust and the like—occur in broken or obtuse-angled crystals which have a comparatively poor cutting action and tend to roll over between lapping surfaces, so that they become sub-angular or rounded in the process of wear, and may in some circumstances practically cease to have any cutting action at all. While this limits their usefulness for general purposes to such an extent that they are scarcely regarded as practical abrasives in engineering practice, there are occasions when a mild and very rapidly diminishing action is useful, especially in final lapping operations between mating parts. Very finely divided Cornish silica, for instance, is sometimes used for lapping sliding instrument parts, or nuts and screws, in cases where an extremely free and silky motion is desired, as in the focussing slides and screws of microscopes.

Diamond dust is extensively used for lapping very hard metals, particularly on rotating laps, which are in some cases used in the same way as grinding wheels for finishing the edges of tungsten carbide tools. It has been found that lapping produces the highest possible finish, and thus the maximum resistance to initial wear of the cutting edge, which lasts much longer than if finished by ordinary grinding.

### Honing

This term refers, in its literal and original sense, to the sharpening of edged tools by rubbing them down by hand on a " hone " or slip of abrasive stone. In modern engineering, however, it has come to represent a development of the process of lapping, in which the charged metal lap is replaced by an abrasive slip. This results in much faster and more efficient action than ordinary lapping, with improved convenience of operation by eliminating the trouble and messiness of using abrasive paste on the lap.

Honing is usually done by means of one or more small abrasive slips fitted to a tool which, in the case of internal

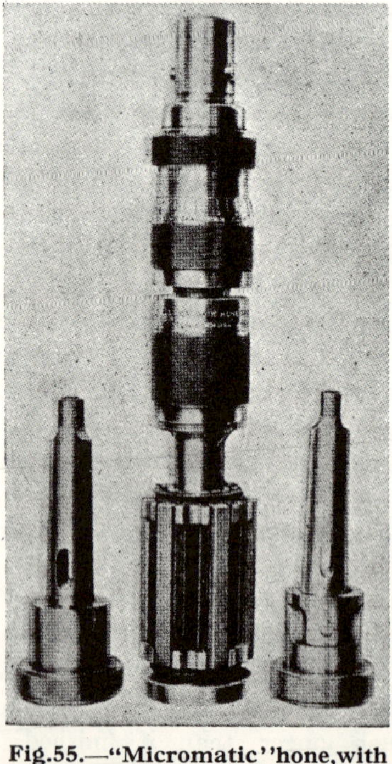

**Fig.55.—"Micromatic" hone, with rigid and floating driving arbors**

application, resembles an expanding reamer, and like the latter, is capable of diametral adjustment. In the " Micromatic " hone shown in Fig. 55, a number of narrow slips of artificial abrasive material, mounted in metal holders, are fitted to the hub spider in such a way that they can be expanded radially by suitable adjusting gear. Some forms of hones have a smaller number of abrasive slips, alternated with metal or fibre guide pads, which also have radial movement, and ensure that the tool generates a truly circular hole, irrespective of initial error in the roundness of the bore.

Other kinds of honing appliances have only one abrasive slip, in conjunction with two guide pads, only the slip being adjustable to fit the cylinder bore. These are commonly used for comparatively small bores, down to about $\frac{5}{8}$ in. dia. ;

**Fig. 56.—The Micromatic Hydrohoner, for automatic honing of small bores**

nearly all the larger honing tools have at least three slips. The latter are readily removable, so that different grades of absasive may be used successively on a single bore, if desired. The means of expanding the slips may include some form of automatic feeding device, or a method of adjustment which can be operated without stopping the rotation of the tool.

## Honing Machines

Appliances as above described are used in almost exactly the same way as laps, and some forms of them may be applied by means of a lathe or drilling machine. Some kinds of honing appliances used for reconditioning motor car cylinders are driven by a freely floating shaft from any convenient source of power, such as a line countershaft or an electric hand drill.

There, are, however, many advantages in using specially designed machines for the honing process, and this course is almost exclusively adopted in production practice, not only for convenience and expedition, but also because such machines can be made to carry out their work automatically, thus eliminating both the need for individual supervision and the element of human skill.

One of the most popular machines for honing small parts is the Micromatic Hydrohoner (Fig. 56), which is made in single- and double-spindle types, with a capacity from $\frac{1}{4}$ in. to 1 in. diameter, and $\frac{1}{4}$ in. to 6 in. length. The spindle is horizontal, and besides having a rotary movement, is also reciprocated by means of an adjustable swashplate mechanism. It is important to note, however, that the reciprocating and rotating movements are not in exact phase, but are definitely put into a slightly different relationship at each stroke. The action is analogous in its effect to that of the machines which are used for winding duolateral or "honeycomb" coils, the object, of course, being to ensure that each individual abrasive grain takes a slightly different path relative to the bore surface at each stroke.

The result is the most perfect evenness of action and avoidance of any possibility of scoring.

The work is mounted in a fixture on a table facing the spindle, and in parallel alignment with it. Axial movement of the table slide is provided by a hydraulic piston located beneath it, and controlled, in respect of limits of stroke at either end, by adjustable stops on the slide, actuating a two-way control valve. The motion of the table feed is relatively slow, and furnishes yet a third factor in the rotary-reciprocating motion of the hone relative to the work. Expansion of the hone, up to a pre-set maximum stop, is automatic, and it is also kept lubricated when in use. The rate at which metal may be removed is up to or over .001 in. per minute, using a hone of 80 to 180 grit, which gives a finish sufficiently high for nearly all commercial purposes, but still finer finish can be obtained by " mirror honing " with stones of specially selected grit, down to about 300 to 500.

Larger honing machines, of both vertical and horizontal types, are used for a very wide variety of production work, including the finishing of long cylinder barrels for hydraulic gear, recoil mechanism of guns, cylinders for steam and I.C. engines, and so on. A comparatively recent development is the external honing of piston rods, plungers and rams, which is usually done on horizontal machines, some of which have exceptional length of bed.

The Barnes Drill Co., of Rockford, Illinois, U.S.A., are specially noted for the production of large honing machines, an example of which is seen in Fig. 57. A special feature of these machines is complete automatic operation, including compensation for errors which might be liable to occur in blind-ended holes, and the hone may be set to withdraw and stop when the required dimensional limit has been reached.

It has been found that the angle of helix produced by the rotary-reciprocating motion of the hone has an important bearing upon the smoothness and accuracy of the work. The included angle between the helixes of strokes made in opposite directions should be between 40 and 60 degrees, which means to say that the stroke of the hone per revolution should be from about 1 to $1\frac{1}{2}$ times the diameter of the bore. Honing speeds recommended for cast-iron range from 200 to 250 f.p.m. (peripheral speed of rotation) and from 50 to 75 f.p.m. (linear speed of reciprocation); the respective figures for steel are 150 to 200 f.p.m. and up to 40 f.p.m.

### Comparisons between Lapping and Honing

Lapping and honing cannot be regarded as identical in their action, though their technique and general effects are similar. Honing is a more definite *cutting* process, and the particles of abrasive being positively held in place in the hone, the action is quicker and maintained at full efficiency throughout the operation, with much less tendency for them to be transferred and become imbedded in the surface being worked upon. It may be said that, at least from the production point of view, honing is a more practical and useful process than lapping; but for special operations in the tool room, or in circumstances where elaborate and specialised equipment is lacking, highly accurate work can be done very simply by lapping.

In both these processes, internal and external surfaces may be finished to extremely close limits of roundness, parallelism and straightness, but neither the position nor the alignment of the axis can be established or affected; the lap or hone will always follow the general direction set up by the previous machining operations. In neither case should the tool be subject to any side restraint, but must be free to "float" so that it can find its own alignment; if this rule is not observed, cylindrical and parallel accuracies are bound to be affected

### Standards of Finish

The realisation of the importance of really high finish on

wearing parts of machinery, and the introduction of special methods of obtaining such finish, have made it necessary to institute standards of finish and methods of measuring it. Simply judging the finish by eye has no real evidential value, neither does it enable any definite and universal standards of comparison to be established.

Any highly finished metal surface acts as a reflector of light, and its efficiency in this respect is proportional to the fineness of the finish, so that it is possible to obtain some measurement of finish by photometric methods. In the instrument used for this purpose, a ray of light of known intensity, directed on to the surface of the specimen, produces a reflected ray, the strength of which is measured by a photoelectric cell. While this method is quite a useful and practical one for certain purposes, it is not universally applicable, especially in view of the fact that different materials differ widely

**Figc 57.—Automatic verti(al honing machine** *Barnes Drill Company*).

in their reflective efficiency, apart from the matter of finish.

The instrument most widely used at present for measuring finish is the Profilometer, which works on a purely mechanical principle, incorporating a minute gauge point which moves over the specimen (or under which the latter is passed), and which records microscopic deviations or roughness by means of an amplifying system. The directness and comparative simplicity of this method of measurement are important practical advantages, especially as it readily discriminates between generated accuracy and mere polish, which might be obtained on an inaccurate surface by buffing. It will detect and measure irregularities as small as one millionth of an inch in finished surfaces, and standards of finish are therefore estimated in increments of this amount, or in other words, micro-inches. A surface finished within limits of 3 to 5 micro-inches is smooth enough for most commercial purposes, but much finer finishes are often obtained on parts such as Diesel engine fuel pump barrels, by mirror honing.

The illustrations of honing equipment in this chapter are reproduced by courtesy of Messrs. Gaston E. Marbaix, Ltd., the British agent for Micromatic and Barnes Honing Machines.